A Letter to Dabjebrosato

Are You an 'Intelligent Animal' or a 'Created Human Being'? Find the Answer to This Question and You Will Find Love, Peace and Security

J A Moore

A Letter to Dabjebrosato
Copyright © 2021 by J A Moore

All rights reserved. No part of this publication may be reproduced, distributed, or transmitted in any form or by any means, including photocopying, recording, or other electronic or mechanical methods, without the prior written permission of the author, except in the case of brief quotations embodied in critical reviews and certain other non-commercial uses permitted by copyright law.

Tellwell Talent
www.tellwell.ca

ISBN
978-0-2288-5159-2 (Hardcover)
978-0-2288-5158-5 (Paperback)
978-0-2288-5160-8 (eBook)

CONTENTS

Reviews v

Dedication vii

Introduction ix

Chapter 1 Who are you and me? And what is our place in the World? 1

Chapter 2 How to research whether a *Scientific Theory* is an established scientific fact? Or is it just a fantasy? ... 4

Chapter 3 Researching whether the *Theory of Evolution* is an established scientific fact? Or is it just a fantasy? ... 6

Chapter 4 What is 'The Theory of Evolution'? 12

Chapter 5 What is 'The Big Bang Theory'? 13

Chapter 6 What is the 'Primordial Soup Theory'? 15

Chapter 7 Do the 'Evolutionary Theories' stand up to common sense scrutiny? 16

Chapter 8 Where are we now? 20

Chapter 9	Researching whether the answer *'Created Human Being'* is scientifically possible and does it pass the common sense test?..22
Chapter 10	Who is this Omnipotent Creator Being?.............29
Chapter 11	The Bible is the Basis of Christianity – Is it archaeologically, and historically accurate and reliable?..33
Chapter 12	Is Jesus an actual historical figure, who lived in the Middle East 2000 years ago?.........................35
Chapter 13	Was Jesus actually killed by being crucified (nailed) to a cross?..37
Chapter 14	Did Jesus come alive again, after his body was placed in a burial chamber and was he seen (physically) by five hundred people afterwards?...39
Chapter 15	What are the characteristics of the Christian God and Jesus?...42
Chapter 16	Why did Jesus crucifixion become the greatest gift ever? 46
Chapter 17	What does all this, mean for us today?................54

Endnotes..59

REVIEWS

I have finished reading A Letter to Dabjebrosato and must say I have really enjoyed it. What a great way to engage your Grandchildren and I am sure many folk would purchase it as a way to engage with their children. I loved the thought-provoking thread of questions that maintained my interest and the simple, common sense answers that gave complete answers.

David Beckwith. Grandfather to thirteen.

Cheryl and I have read A Letter to Dabjebrosato and we both agree it an excellent resource. I love the concept, of a Grandfather teaching his Grandchildren. It would be an excellent gift for any parent, grandparent, or friend to give to a loved one who is not a Christian.

Rob and Cheryl Bridgfoot. Grandparents to six.

Reading "A Letter to Dabjebrosato" has illustrated how we can encourage our grandchildren to wrestle with the important issues of life and think about who they put their trust in, to steer them through an increasingly challenging world in which we live. We would gladly share it with our own grandchildren as they reach their teenage years.

David and Jenny Price, Grandparents to ten.

In these pages you will feel the spirit of my friend J. A Moore, his intimacy with God, his discipline as a scientist and his passion for his grandchildren. The scientific-like presentation on the source and meaning of life, man's fall from God's order and God's subsequent redeeming action through Christ. Is the product of J. A's life experiences, study and contemplation on the Word.

John Gardiner, Christian, scientist and grandparent of eight.

DEDICATION

This book is dedicated to my children and grandchildren, they are very important to my wife and I. My main aim is for them to learn, how to have the best that life can offer, this was the inspiration that continually encouraged me to research and write the book.

The book is also dedicated to every child and grandchild worldwide, with the aim that they might learn, how to have the best life possible, as without them we have no future!

INTRODUCTION.

My Dear Dabjebrosato

Dear Dabjebrosato,

Upon reading this letter, I imagine that the first question that most likely will come to your mind will be, who is Dabjebrosato?

Well it is you my *special* Grandchild. The name is made up from the letters of the names of my grandchildren, namely (the late David), Aidan, Ben, Jess, Brooke, Sarah, and Tom. Each of you are a very important grandchild to Nan and I and we very much want each of you to have the most, satisfied and fulfilling lives that you can.

At the same time, as what I will be discussing with you affects, everyone in the World, irrespective of race, sex or religion, if anyone wants to read along with us and join in the discussion, I would be pleased for them to do so.

To set the scene. From time to time I have a dream which distresses me greatly.

Nan and I are standing just inside the gates of Heaven, waiting for each of you to come, as you will when your allotted time is complete.

We have gone before when our allotted time was up. And with what we knew and had been taught, we had been able to book our place in Heaven. (We were looking forward to it at our age) and knowing the passwords, entering was no problem.

However what has been worrying me in my dream, is that each of you have not been taught, who you are or what place you have in the World? And more importantly, you have not been taught the fact that Heaven exists, the need to make a booking if you are to go there (and not the other unmentionable place) and the need to know the password to enter it.

So in my dream, I see each of you arrive with so much expectation, only to be turned away because you haven't learnt of the need to make a booking and what the password is. It causes you so much grief and tears. And it breaks Nan and my hearts as we stand there helplessly, watching you turned away, hearing your cries of anguish, and shouts of anger. Saying, "Why did no-one tell me, this might happen?" As you fade away out of our sight.

This brings me to the importance of my writing this letter, that you by reading it, may learn what you have to do, to avoid the terrible scenes in my dream from happening.

I would like you to know, that quite a large part of what follows will be scientific discussion, but in an everyday sense that I am sure you will understand. I would encourage you to persevere, as the rewards for doing so will benefit you for the rest of your life.

Let us begin.

CHAPTER 1.

Who are you and me? And what is our place in the World?

This is an extremely important question that you and I will discuss together. as I know that the answer we find to the question, will affect how we approach every aspect of our life.

The question is centred on this, "Are we just *'Intelligent Animals'*? Evolving from nothing, living with no meaning or purpose, in a Universe and World with no meaning and purpose and when we die, returning to nothing?"

Or are we *"Created Human Beings?"* With an earthly body and an immortal soul, created by a supreme being, born into a Universe and World with a meaning and purpose. And when we die, while the earthly body shell remains, the immortal soul moves to a different dimension where it may face judgement as to how we have lived, depending on the choices we have made"

To me, the answer we can provide to this question will affect, all the decisions we make and how we live our lives for every moment, of every day.

The first answer that we are just *'Intelligent Animals'* is the one some people give *because they are Atheists, who believe there is no God*. And because they believe there is no God *(or Creator/Designer)*, they believe the Universe, the World and you have evolved by the 'Theory of Evolution' from nothing! All of it (your life included) has no meaning or purpose, and when you die, your body and brain will disintegrate to nothing and you will no longer exist.

In short, you are only here for a short time, so you need to live life to the full, selfishly grabbing and getting as much as you can, before it ends.

The other answer that we are *'Created Human Beings'*, is the one that is given by some people *(Christians, who believe in a creator God)*. They believe that the Universe, the World, and you are not here by accident! That it has all been created by a supreme being God, *(You might better understand 'God' as an architect, artist, weaver, potter, only on a scale that we can't fully comprehend)*.

This belief means that because you are created, you are a unique and special individual, with a life that has a purpose, (that can be searched for and found what it is). That you have an earthly body, brain and spirit, and an immortal soul (that means it lasts forever). While your body, brain and spirit will disintegrate at death, your immortal soul will enter another dimension, which will be either Heaven or Hell, depending on the choices you have made. (Note that *Heaven and Hell* are the result of religious belief, for Atheists, there are no such places).

This variance makes who is right a very important answer.

First we will look closely at the claim that we are just *'Intelligent Animals'* and everything is the result of 'The Theory of Evolution'.

But before we do, we must work out how to research, whether a *Scientific Theory* is an established scientific fact? Or is it just a fantasy?

CHAPTER 2.

How to research whether a *Scientific Theory* is an established scientific fact? Or is it just a fantasy?

To you Dabjebrosato, I am writing to you as a Grandfather, who by experience knows that as you enter the adult world, you will be pressured to accept many different ideas and propositions as established facts. It is important that you don't accept these ideas at face value, but study and investigate them closely to see whether they are true or not.

Before we begin, I must stress that when we are judging ideas and propositions, we should only focus on those issues and propositions and not on the people proposing them. After all generally the people concerned are only repeating what they have been taught and particularly on some big issues peer pressure can be very forceful and even if a person has a strong feeling that what they believe is not correct, they are unwilling to say so because of the risk of being rubbished or worse by their peers.

I also would like you to understand two things. The first is that truth is an unchangeable basis on which to judge something *(Truth definition: A statement proven to be or accepted as true)*:[1] If what is being said, can be shown to be not based on researched truth. Then irrespective of how well it is promoted or what great claims are made about it, if it is not true, then it is just a fraud.

The second is in relation to science, *(Science definition: The observation, identification, description, experimental investigation, and theoretical explanation of phenomena:*[2] What I would like to emphasise here is that science has only one role and that is of discovery.

Discovering the how and why something is like what it is. Discovering what happens when you mix this liquid with that one. Discovering what reaction you get when a specific force is applied to an object. Discovering what ingredients are needed to be put together to make, for instance 'steel' or a 'pizza'. Discovering the laws of aerodynamics and what components are needed to be put together to make an aeroplane that will fly.

One thing Science is not and cannot be! Is a creator. Science never has and never will, create something out of nothing!

CHAPTER 3.

Researching whether the *Theory of Evolution* is an established scientific fact? Or is it just a fantasy?

Darwin's *'Theory of Evolution.'* *'The Big Bang Theory'* you hear of these theories, in classrooms, Universities and in general conversation. But what is *"The Theory of Evolution?"* And is it based on scientific facts or is it just a fantasy in someone's mind?

As we look at the issue of The Theory of Evolution. "I would like you to be more than just a reader, I would like you to work with me as a co-researcher, seeking answers to the questions I'm asking and thinking of other questions that come to your mind. This is because the results of what we will be researching, will have a major effect on how you see things in the future. I also want to equip you, to be able in the future to challenge what people might say is factual, by putting them on the spot and asking for supporting evidence that establishes what they are saying can be substantiated as factual."

I would say, that today the *'Theory of Evolution.'* is the most important scientific theory that is being taught in Schools, Universities as an established fact. This scientific fact we are told has been established by scientific research, study and reason and is accepted, by most of the population around the World without question.

Down through the ages, the story of Creation as it is told in the book of 'Genesis' in the Bible, was believed to be the way the World and the Universe came into being. However during the period of 'The Enlightenment'[3] (1700's) the theories of the World coming into being by 'The Theory of Evolution' came into prominence. It appears to me, that this was a way of providing an opposing view to the Biblical explanation.

To me there are two ways of looking at a subject. The first is in a scientific way only, looking at what is around us in the material World and Universe, then establishing a theory. Which is (theory definition: A set of statements or principles devised to explain a group of facts or phenomena, especially one that has been repeatedly tested or is widely accepted and can be used to make predictions about natural phenomena.)[4] and believing that it is the only way the World we see around us, could have come into being.

The second is by using scientific research but with added common sense, not believing everything we are told, but asking is it possible that what we are looking at could have happened as a result of that theory? And if it is not possible, what then is the best explanation of how it came into being.

As we look at what is proposed, we need to be asking three questions, to test that what is being proposed has happened, in the way those who support that theory say it came into being?

The mathematically possible question?

The first question we need to ask is: "Is it mathematically possible to have happened as proposed by the *'Theory* of Evolution'?"

When I talk about something being mathematically possible, what I am talking about is calculating the times that a certain result is likely to happen, when it is only happening by random chance.[5] A very relevant illustration is the mathematical calculation of the chances of a person picking the winning 6 numbers in a lottery, where there are 49 numbers to choose from. It has been calculated that the chance is 1 in 13.98 million. (That is you would need to buy 13 million, 983 thousand and 816 tickets to win just once.) When you increase the numbers to choose from 49 to 59, the chance of picking the winning 6 numbers increases to 1 in 45 million.[6]"

As we will see when we look closer at the Theory of Evolution, (which claims that everything that occurs in the world, just happens by random chance), it becomes a gigantic lottery, with thousands of balls. This is because the world has millions of incidents/events happening every moment, such as "babies being conceived, flowers flowering and being fertilized, snow falling, each snowflake being non identical". Applying mathematically possibilities to what is happening in the world, as happening by evolution would give you a figure of 1 chance in 10,340,000,000.[7].

Looking at this figure with common sense, I believe we can say with almost certainty, that it is impossible for the Universe, the World, yourself and myself to have evolved from nothing to where it is today.

The irreducible complexity question?

The second question, (Also known as the irreducible complexity question) is: Are all the necessary parts in place for the entity to function? The most well-known of these irreducible complex functions is the one given by: *Creationist Michael Behe'. His illustration is of a mousetrap. A mousetrap consists of a number of parts. There are: (1) a flat wooden platform to act as a base; (2) a metal hammer, which does the actual job of crushing the little mouse; (3) a wire spring with extended ends to press against the platform and the hammer when the trap is charged; (4) a sensitive catch which releases when slight pressure is applied; and (5) a metal bar which holds the hammer back when the trap is charged and connects to the catch. There are also assorted staples and screws to hold the system together. If any one of the components of the mousetrap (the base, hammer, spring, catch, or holding bar) is removed, then the trap does not function. In other words, the simple little mousetrap has no ability to trap a mouse until several separate parts are all assembled.*[8]

This irreducible complex question can be applied to millions, if not billions of separate organisms, and activities in our World. Take for instance the relationship between trees and flowers and bees. If a tree had evolved singly without another to pollinate with, then it could not reproduce. Even if there were two trees, without flowers and bees (both themselves irreducible complex entities) then the trees could not pollinate each other.

Another instance that each of us is very familiar with, is the human body. An organism that evolved, with a mouth alone, (no tongue, no teeth?) without a throat, stomach, small intestine, large intestine, and anus, could not function and would eventually die. Or thinking further it wouldn't be alive in the first place, would it?. *"You can think of other scenarios, I am sure; you will find the possibilities are endless."*

Could the 'spark of life' have happened spontaneously?

Which leads to the third question we need to ask which is: "Could the organism, human, animal, plant, whatever else you think of. How could it have developed the spark of life, just spontaneously? E.g. The heart just starting to beat. We know from medical studies that a heart is driven by electrical impulses. To start a heart that has gone into cardiac arrest, requires a sudden shock of electrical stimulation. Where did a heart that had evolved, get this kick start? Perhaps it was hit by lightning? But then would you be left with a beating heart or a barbequed mess?

And remember this action is not required to happen just once, but with each blood circulatory system, respiratory system, digestive system, etc of one human and all the systems must be started at one time. Multiply that by billions of humans, add in dogs, cats, all animals, fish, insects. *"I will let you add to the list! But you can see the mathematic possibility of this happening by accident is astronomical."*

Looking at the scenarios we found in questions number two, the answer is unless the organism, human, animal, plant, had happened to be formed in a basic completeness in an environment of basic completeness then the organism, human, animal, plant would not have evolved into anything and if the kickstart of life in question three, hadn't happened then the subject would be lifeless. These are essential questions, as it doesn't matter how good the theory sounds, if the subject matter dies or doesn't live in the first place, it's Game Over!!!!.

"Dabjebrosato, could I suggest that at this stage, it would be a good idea for you to get a nice drink, sit down and put your thinking cap on. Because we are going to have a detailed look at the basis of the Theory of Evolution and that is going to stretch your mind and your sense of the incredulous, as I found when I researched and started to put this book together.

And at any time if you want to ask me a question, of anything you do not understand, please do so. As it is only by discussing these issues can we fully understand them and what they mean for our lives.

Could I also encourage other readers, to do the same, my idea in writing this book is to set out the issues as I see them. Researching and discussing them with others, will allow you to see whether they are true or not.

CHAPTER 4.

What is 'The Theory of Evolution'?

The theory of evolution by natural selection, was first formulated in Charles Darwin's book "On the Origin of Species" in 1859.[9] *The Theory of Evolution* as I understand it. Is that everything we see (and can't see) in the World and the Universe began out of a void, emptiness, nothing. All of it came into being, through a process of continuous evolving, that took millions of years and involved billions of major and minute mutations and changes.

First it *'just happened'*. (Note: You will find that *'just happened'* is a very vague comment that is used a lot when explaining *The Theory of Evolution*). It *'just happened'* that there came into being, the physics that make up the *'Laws of Physics'*. These laws encompass Newton's Laws of Motion and other physics relating to gravity, heat, electricity, magnetism and light. These *'Laws of Physics'* hold everything in the Universe together and regulate and coordinate how it all functions together, so that works like a billion clocks all keeping their own time, but at the same time coordinated together so the Universe functions as one mammoth timepiece.

CHAPTER 5.
What is 'The Big Bang Theory'?

Another part of the puzzle is *'The Big Bang Theory'*. It is explained that: *'Our universe is thought to have begun as an infinitesimally small, infinitely hot, infinitely dense, something, a singularity. After its initial appearance, it apparently inflated (the "Big Bang"), expanded and cooled, going from very, very small and very, very hot, to the size and temperature of our current universe.*[10] It continues to expand and cool to this day and as it cooled it amazingly developed, galaxies, stars, planets and all that makes up the Universe.

As the Universe took shape in one galaxy, in one solar system, the planet Earth, along with its Moon just happened to come into being as an extremely hot, ball of gas, dust and other material.[11] This ball of 'whatever' was spinning and again just happened to lock into an orbit around our Sun, where it completes an orbit every 365.256 days or one year.[12] As well the Earth's spinning rotation, just happened to settle at one rotation each 24 hours or one day.

And the amazing coincidences are not finished either, it's rotating on an axis that just happened to be 23.5degrees off vertical.[13] It is this tilt of the axis that gives us the seasons, Autumn, Winter, Spring and Summer and makes up the variety of temperatures, rainfall, etc that allows everything (living and static) to go through their life cycle.

At this time remember that when the Earth originally formed it was incredibly hot (the temperature was thought to be millions of degrees Celsius[14]) this needs to be kept in mind for future discussion, as it raises the critical question. Is it possible for life to exist at such a temperature, to later appear when the Earth's surface cools?

CHAPTER 6.
What is the 'Primordial Soup Theory'?

As it cooled, the next evolutionary theory is that on this planet Earth, there came to be a primordial soup. *"Primordial Soup" is a term introduced by the Soviet biologist Alexander Oparin. In 1924, he proposed a theory of the origin of life on Earth through the transformation, during the gradual chemical evolution of particles that contain carbon in the primordial soup.*[15] From this primordial soup emerged a single cell, which then divided into two cells. Which then evolved and mutated and evolved, over billions of years, until we have what we see around us today. The culmination of this evolution is humankind. With a brain more complex than the best of computers (with much of the complexity which is still unknown.) With the ability to reason and to remember in a way that is far in advance of any other species.

That broadly completes the Evolutionary theories that the 'Theory of Evolution' is based on, and explains in more detail, what those who believe in the theory, think is how the Universe, the World and you evolved to what we see around us today!

CHAPTER 7.

Do the 'Evolutionary Theories' stand up to common sense scrutiny?

Now we can look closely at these theories and see if they stand up to close examination and the application of common sense?

The Laws of Physics.

Firstly could the 'Laws of Physics' have just evolved from nothing? These 'Laws' such as 'Law of gravity', are vitally important to be in place, in order that the stars, suns, galaxies, planets, everything in the Universe stay in exactly in the right place. As well they must be extremely, accurately, co-ordinated. If the gravity of the Sun was a fraction stronger than it is, then it would have drawn the Earth in closer and it would have got hotter and then burnt up, or if it was weaker the Earth would have drifted into outer space and become a frozen wasteland.

Some other *'Laws of Physics'* are 'Newton's Three Laws of Motion', 'Conservation of Mass-Energy', 'Conservation of Momentum', and 'Laws of Thermodynamics'.[16] All play a very important part in the ability of the Universe to function as we see it before our eyes in everyday life and through electron microscopes and space telescopes.

For the Evolutionary Theorist, these *'Laws of Physics'* just happened to be available in the empty space of the Universe waiting for the 'Big Bang' to occur. "Really?" To me that answer doesn't pass the common sense test. Common sense tells me that for the *'Laws of Physics'* to be in place there had to have been some force/thing to have created them. *To cause to exist; bring into being:* [17]

The Big Bang Theory.

Next, we can look at the 'Big Bang Theory' from a common sense point of view. With your and my experience of what we see when an explosion takes place, (after all it is called "The Big Bang Theory") all I would expect to find would be massive destruction? With particles of debris flying in random directions everywhere? But that is not what we see at all. How is it then that instead of chaos, we see such amazingly different objects (solar systems made up of suns and planets interacting with each other in very, very intricate ways? Again, just like the workings of a huge clock come to mind.

For instance, in one small way, to understand this intricacy, we can look at our Moon and the way it interacts with the Earth.

The evolutionists have an 'evolutionary' theory, to explain how our Moon came to be where it is today, which is called the 'The standard giant-impact hypothesis'. This suggests a Mars-sized body called Theia impacted Earth, creating a large debris ring around Earth, which then accreted to form the Moon'.[18]

But when we look at the mathematics involved in the precision of the orbit of the Moon around the Earth, the idea that it just came into being by the accumulation of debris from a collision would seem impossible to believe.

For instance every time the Moon orbits the Earth, it is exactly a time period of 27.322 days.[19] *Every orbit is the same time exactly, not a second more or less. As well every orbit is elliptical so that the Moon rises (known as the Apogee) to a highest distance from the Earth and then descends to a lowest point above the Earth (known as the Perigee). While these distances vary from orbit to orbit, the elliptic of the orbit is so precise, that the Apogee and Perigee can be calculated to the day, hour and minute, for tens of years in advance. But there is more. The orbit of the Moon is not parallel to the Equator. but is in a diagonal orbit across the Equator from North to South, so that in each orbit the Moon rises above and falls below the Equator. But while the Monthly movement of the orbit above and below the Equator varies gradually over an eighteen-year period, amazingly for nine years it gradually rises and falls until it settles at a level of rising and falling that is exactly 28.45 degrees above or below the Equator. Then for the next nine years the orbit gradually decreases until it is only rising and falling at a level of 18.09 degrees above or below the Equator. Then the cycle begins again. All the while the movement of the Moon and its gravitational pull has an effect on the Earth's surface such as controlling the ocean tides and weather systems. Plus, there is much more that scientists haven't discovered yet.*

The Primordial Soup Theory.

The third evolutionary theory that we have to put to the common sense test, is the *theory that life on Earth evolved from a 'primordial soup'?* So we need to consider the major problem that arises with the intense heat that the Earth was subjected to following 'The Big Bang'. If the Earth was a big ball of gas, dust and other material, at a temperature of millions of degrees Fahrenheit, could it be possible that life which today can only exist at a temperature that ranges between roughly 40 degrees Fahrenheit *(4.44 Celsius)* and 108 degrees Fahrenheit *(42 Celsius).*[20] Could have existed during those extreme temperatures, only *"to come to life"* when the earth cooled to the temperatures we experience today? Using common sense and my experience, I believe that would be impossible and I would expect that you would agree.

The DNA Dilemma.

Finally the predicament that 'Evolutionists' cannot find an answer too is DNA. (*DNA, or deoxyribonucleic acid, is the hereditary material in humans and almost all other organisms. Nearly every cell in a person's body has the same DNA.*[21] *And without DNA, living organisms could not grow. Further, plants could not divide by mitosis, and animals could not exchange genes through meiosis. Most cells simply wouldn't be cells without DNA.*[22]

And the dilemma for 'Evolutionists' is that DNA cannot be created by the 'evolutionary process' starting with nothing.

Bill Gates (who with Paul Allen founded 'Microsoft' in 1975[23]) said "DNA is like a computer program but far, far more advanced than any software ever created." [24] And like a computer program unless it is perfect (without one gene out of place) it will not create the cell that it is meant to create. And having the *'hallmark' of irreducible complexity'* common sense tells us that like a computer program, DNA cannot evolve from nothing, and therefore it has to have had a programmer or designer.

CHAPTER 8.

Where are we now?

Well, Dabjebrosato? What conclusion have you reached now that we have looked closely with common sense at these evolutionary theories? For my part, from what we have found, I think it is impossible for the Universe, the World, yourself and myself to have become what it is today, from nothing, just by billions of chances as suggested by *The Theory of Evolution*.

And so I believe that we can only conclude that the *Theory of Evolution*, along with the *Big Bang Theory* and *The Primordial Soup Theory* are not based on fact, but are just *wishful thinking* that collapse when subjected to the harsh glare of mathematical and scientific study, coupled with common sense.

In fact, Mr Charles Darwin made A BIG MISTAKE! Do you not agree?

Where does that leave us now? Well can I say in a much better place than if the various *Theories of Evolution* were true. A strange thing to say, you may think, but consider that if the *Theories of Evolution* were true and the World, Universe, you and I had evolved from nothing and everything had just come together by the process of mindless chance, then the universe and world would have no meaning or purpose. And even worse our lives would have no meaning or purpose either. This I believe is why, in the world there is so much depression, anxiety, consumption of alcohol, drugs, gambling, promiscuity, even death *(by your own hand)* is welcomed, hoping that will bring the misery of life to an end in oblivion! People preferring death, because they have no hope for the future.

But if there is an alternative? That offers hope, love, reason and meaning for our lives even during hardship or various difficult times. And not just in the lives we are living today, but also in the life we would continue living *'spiritually'* after death.

Wouldn't that be worth searching for by all means possible?

In the Bible, Jesus told a parable (story) emphasising how important he thought finding meaning in life should be. He said, [44] *"The kingdom of heaven is like treasure hidden in a field. When a man found it, he hid it again, and then in his joy went and sold all he had and bought that field."* [45] *"Again, the kingdom of heaven is like a merchant looking for fine pearls.* [46] *When he found one of great value, he went away and sold everything he had and bought it". Matthew 13:44-46.* [25]

Dabjebrosato, let us search using science and common sense to see if we can find if it is possible that we really are *'Created Human Beings'* and what that would mean for us.

CHAPTER 9.

Researching whether the answer *'Created Human Being'* is scientifically possible and does it pass the common sense test?

As I said previously the alternate answer given by some people *(Christians, who believe in God)*. Is that they believe we are *"Created Human Beings,'* They believe that what is written in their holy book the Bible, explains in some detail, that the Universe, the World, and you are not here by accident! That it has all been created by a Supreme Being *(a creator God)*.

Is this true? let us look at the subject closely, (as we did with the *'Theory of Evolution')* and I believe that the best subject for us to look at first, is our own bodies.

Our own bodies – Could they only have been created?

Dabjebrosato have you ever looked at your own body? Very, very closely, that is. Take a moment, no take a long while. You will find it consists of the most amazing intricate items, that you ever could imagine.

It's difficult to know where to start, every part, every cell has its own fascinating story, it's all mind boggling.

To begin let's look at our eyes for example, they are the most intricate of camera's that you could ever buy. Our eyes self-adjust their exposure, for all sorts of lighting. They automatically focus (10 times a second), on objects near and far, they have a cover (the eyelid) that quickly protects the eye from foreign objects, while also 'blinking' several times a minute cleaning and moisturising the eyeball. And that is only a little of the detail that occurs every second you are awake.

Next we can look at our bodies' blood vessels. *Of which there are three different types, arteries, veins and capillaries. In an adult human these three types combined add up to 100,000 miles (160,934 kilometres) of blood vessels.*[26] *(A distance which would circle our planet four times.) One purpose of the blood vessels is to absorb and take oxygen, from the air which is inhaled into the lungs as we breath and it is circulated around and released into the body to be used by the cells. At the same time a similar volume of carbon dioxide (created as the cells 'burn' the oxygen) is removed from the cells and returned to the lungs to be released back into the atmosphere.*[27]

That's just two examples, as well there is the nose, ears, hands, fingers, feet, bones *(270 at birth)*, skin, and the brain, *(which is capable of more functions than the greatest computer, ever made by man)*, and there are hundreds more, each of which have their own extremely, intricate, complexities.

The mathematically possible question?

Now as we did with the *'Theory of Evolution'*, we must ask question one. But this time the question is framed as "Is it mathematically possible to have happened by *'an omnipotent creator God*?" I use the word *omnipotent* because it means 'having unlimited or universal power, authority, or force; all-powerful.[28]

And while it takes a while to get your head around the idea of, *'an omnipotent creator God'* it does allow the Christian to answer, yes, to the question. Yes, it is mathematically possible for *'an omnipotent creator God'* to have created the Universe, the World and you. Because *'an omnipotent creator God'* can do things beyond our human comprehension and wouldn't be God, if it could be comprehended.

We could get side tracked into whether *'an omnipotent creator God' can scientifically be proven to* exist. But we'll leave that debate for another time and continue on the basis that the Christian believes that *'an omnipotent creator God'* exists by *'faith'*. *Faith Christianity A secure belief in God and a trusting acceptance of God's will viewed as a theological virtue.*[29]

And while you might think that only Christians have to have faith, to believe in *'an omnipotent creator God'*. In fact everybody has *'faith'* in some belief or world view as to how they understand the Universe, the World, and you came into existence. The atheist has faith in the *'Theory of Evolution'*. The naturalist has faith in *'Nature or Gaia'*. As the power source that brought everything into being. *In Greek mythology, Gaea (or Gaia), the primordial earth or mother goddess was one of the deities who governed the universe before the Titans existed* [30]

Stop for a moment and think. Up until now, what was your belief or *'faith'* as to how the Universe, the World, and you, came into existence? Was it as an atheist? A naturalist? Or a Christian?

The irreducible complexity question?

Now let us look at question two. Which we know as the irreducible complexity question, which in common language is, *'are all the necessary parts in place for the entity to function?*

Again we need look no further than at our own bodies. We need to be able to obtain, eat and digest our food in order in to remain alive. Before we even obtain the food, there is the irreducible complexities involved in every bit of food that is available.

The apple tree that produced the apple, required that there be the soil that held the tree upright, that provided the nutriments, the moisture in the form of water, for the tree to grow, the tree roots, the trunk, the bark, the bigger and smaller branches, the leaves (which take time to develop from buds, and go through the process to become mature leaves). Then the tree needs to be programmed, in order that the flowers appear at the right time in the right season, to develop into apples. The bees which pollinate the flowers, have to be present (no bees, no pollination, no apples). All these (each with their own irreducible complexities) needed to have come into existence, in the one instance.

Likewise it is the same with the wheat crop that contained the grains of wheat, that was ground into flour to make the bread. The cow that ate the grass, and processed it into milk, which was milked of, then transported and processed to eventually become cheese.

As I mentioned previously we can then apply the irreducible complexity question directly to our own bodies. We all at the same time, need to have eyes to find the food, legs to get to it, arms with hands and fingers to collect and process it. Then a mouth to eat the food, teeth to chew and grind it, the tongue (whatever it does?), the throat and oesophagus to swallow it, (while at the same time allowing air and oxygen into the lungs and carbon dioxide out, all the while the heart doesn't miss a beat). The stomach and small intestine to digest it, after which the large bowel collects the refuse, before it's discharged via the anus. At the same time the bladder collects the urine and then is able to hold, until it can be disposed of as and when required.

As we found earlier when looking at the "Theory of Evolution" all these individual items, have to be in place at the at the same time, without just one of them the whole body doesn't function or for that matter come into being in the first place.

Having looked closely at this question with common sense, I am of the view that only the Christian view of there being *'an omnipotent creator God' who could create everything in an instant*, can meet the requirements of the "the irreducible complexity' question, To me it is the only plausible answer, as to how these parts came into being in the right place and at the right time for everything to function as it does.

What is your opinion Dabjebrosato think? Can you agree with that? If not what other process can you suggest?

Could the 'spark of life' have happened spontaneously?

Lastly we come to the third question we need to ask: *"Could the organism, human, animal, plant, whatever else you might think of. How could it have developed the spark of life, just spontaneously?*

For this question science enables us to look closely at how life comes into being and the one thing that has amazed me, is that life in any and everything just does not happen spontaneously from nothing, in every generation. Once (at the beginning) the spark of life was planted into every seed or egg, of every living organism, be they mollusc, fish, insect, animal, or human being and then has to be passed from one generation to another or that particular species dies out and becomes extinct.

For instance static items, such as grass and trees, mostly reproduce from a seed. A seed is a dormant embryo—a fertilised egg cell, that formed from cross pollination of two separate plants of the same specie. [31] I would point out to you here that when the seed/embryo was formed, the spark of life was passed from the parent plants to the seed, where it became a dormant form, (in suspension if you will) waiting for the right conditions to germinate (or come to life).

For objects that move, fish, insects, animals, human beings, they require two (one male and one female) of the species, to provide a sperm and an egg, which combine to create a fertilised egg, which can become a new generation of the species. Again the spark of life, needs to be passed from *'the parents'* to the new *'offspring'*.

If for some reason the fertilised egg, seed, deteriorates, or is damaged in some way or the particular species is reduced to having only one plant, or animal then it will become extinct.

On a personal note, one little exercise I have liked to think about since I was a boy is this. For me to be here now, there has been for thousands of years an unbroken chain of beating hearts, (each having a part of my DNA). From the year 1862 onwards through World War I, then the 1929 Depression and World War II it was the hearts of my great grandparents, grandparents and parents, until I was born. For every moment back to the time when Jesus walked on Earth and to the beginning of time a heart was beating with my and your DNA in it.

If at any time during that period had one of those hearts stopped beating, before it had passed the 'spark of life' on to the next generation. Then you and I would not be here.

As I mentioned earlier. Science is not and cannot be a creator. This is particularly the case with the spark of life. You will note, that with invitro fertilization, the specialist cannot make a sperm or an unfertilized egg, which are needed create an embryo. One or both must come from human donors. It's the same with stem cell research, the scientists need human stem cells to work with, because they are not able to make them from nothing. Scientists may be able to make an artificial heart, but it is a very short-term substitute for an actual human heart.

And as with the *irreducible complexity* question, fitting a corpse with an artificial heart, then turning it on, will not bring the corpse back to life.

Well Dabjebrosato, what do you think is the answer is to *'How did the spark of life come into being'*?

For me the answer is that we are not *Intelligent Animals*, that came into being, as the result of the *Theory of Evolution*, but instead *Created Human Beings* who have been created by *an Omnipotent Creator God*. And being *Created Human Beings* raises some very important questions?

Can we have a relationship with (that is talk to and listen to) this *Omnipotent Creator God*?

From the *Omnipotent Creator God* can we learn the purpose for our lives?

If we are *Created Human Beings does* that mean our lives continue after death?

Before we can find the answers to these questions, we *must* search for and see if we can find out who is this *Omnipotent Creator God?*

CHAPTER 10.

Who is this Omnipotent Creator Being?

To find out who is the *omnipotent creator being*, I am going to focus on the religion of *Christianity and its parent religion Judaism*. As I believe it is the only religion, which through its sacred book the Holy Bible, and the life story of Jesus, most comprehensively, physically, philosophically and scientifically, describes the *how, when, who and why*, of the Universe, the World and you, and how you came to be here today as you are.

Before I put the reasons why I believe Christianity is the way to learn about this Creator God. I would like to outline a brief overview of what Christians believe, which will help you to see where everything fits together, like a complete jigsaw puzzle.

To begin with Christians believe there is a Creator God Spirit, that is so vast, huge, timeless, that He is and operates in spheres and ways beyond the comprehension of our tiny mind/computer.

I say He and not It, because Christians believe that their Creator God has feelings as each human father has, (after all as the Creator, He made our DNA, and like a human father, there is a little bit of Him in all of us). The main feeling is one of overwhelming love for the created human beings that He has made. He also has times of feeling jealous, sad, and angry, which happens when his created human beings, rebel against the rules which He has made, (rules which if followed would give them the best life possible). They go their own way, making gods of wealth, fame, materialism, which He knows only leads to disaster for them.

But as a loving Father Creator, when one of His created human beings (which He very much wants to become His adopted children). Realises that they have made the wrong choices and cries out for His help. Because of His love, He instantly forgives them and comes to their aid. But as a loving Father, He also knows children need to be disciplined at times and also needs to go through the difficulties of life experiences, in order to build up their resilience, character and to grow into mature adults.

Christians believe their Creator God has revealed what He is like, through the detail, complexities and beauty of what He has created in the Universe, and the World around us.

He has also revealed Himself through prophets. *A person who speaks by divine inspiration or as the interpreter through whom the will of a god is expressed.*[32] The prophets at His suggestion have spoken and written books, which then were put together in the sixty-six books that make up the Holy Bible. These *prophets* He chose from the Jewish people in the Middle East, thousands of years ago. The prophets spoke what the Creator God put into their minds, and it was then recorded in written form, by scribes and became firstly the Jewish Torah. Then as other prophets spoke and it was recorded, additional books of Jewish history were added after the books of the Torah, until a book was compiled which became known as the Old Testament.

Then Jesus was born into the World in approximately the year, we know as AD. 01. (We celebrate his birth each Christmas). At thirty years of age, he began teaching and preaching, which three years later culminated in him being crucified, (nailed to a wooden cross until he died). His body was then put into a tomb hewn out of a rock, but on the third day, the tomb was found empty, and his followers came to know he had been resurrected. (He had become alive again, physically appearing to them and to a total of five hundred people).

About thirty years after Jesus' death, books began to be written, about his life by his followers (known as Disciples) and one man called Saul. (Saul was once the sworn enemy of Jesus' followers, but then after a meeting an apparition of Jesus on a highway, became his most ardent supporter. The conversion was so profound his name was changed from Saul to Paul the Apostle). After much debate, what became known as the New Testament was compiled with the Old Testament into what we know as the Bible, by St Jerome around A.D. 400.

As I focus on Christianity, I will seek to provide answers that substantiate questions you might have as to:

1. The archaeological and historical accuracy and reliability of the Bible?

2. Is Jesus an actual historical figure, who lived in the Middle East 2000 years ago?

3. Was Jesus actually killed by being crucified (nailed) to a cross?

4. Did Jesus come alive again, after his body was placed in a burial chamber and was he seen (physically) by five hundred people afterwards?

5. What are the characteristics of the Christian God and Jesus?

6. Why was Jesus crucified?

7. What does it mean for us today?

CHAPTER 11.

The Bible is the Basis of Christianity – Is it archaeologically, and historically accurate and reliable?

To begin with the Bible is a book, that stands out above all others. The Bible was written over a span of 1500 years, by forty writers from three continents. Unlike other religious writings, the Bible reads as a factual news account of real events, places, people, and dialogue. Historians and archaeologists have repeatedly confirmed its authenticity. Archaeologists have consistently discovered the names of government officials, kings, cities, and festivals mentioned in the Bible -- sometimes when historians didn't think such people or places existed. For example, the Gospel of John tells of Jesus healing a cripple next to the Pool of Bethesda. The text even describes the five porticoes (walkways) leading to the pool. Scholars didn't think the pool existed, until archaeologists found it forty feet below ground, complete with the five porticoes.

The accuracy of today's Old Testament was confirmed in 1947 when archaeologists found "The Dead Sea Scrolls" along today's West Bank in Israel. "The Dead Sea Scrolls" contained Old Testament scripture dating 1,000 years older than any manuscripts we had. When comparing the manuscripts at hand with these, from 1,000 years earlier, we find agreement 99.5% of the time. And the .5% differences are minor spelling variances and sentence structure that doesn't change the meaning of the sentence.

Regarding the New Testament, it is humanity's most reliable ancient document. All ancient manuscripts were written on papyrus, which didn't have much of a shelf life. So people hand copied originals, to maintain the message and circulate it to others. Few people doubt Plato's writing of "The Republic." It's a classic, written by Plato around 380 B.C. The earliest copies we have of it are dated 900 A.D., which is a 1300-year time lag from when he wrote it. There are only seven copies in existence. Caesar's "Gallic Wars" were written around 100-44 B.C. The copies we have today are dated 1,000 years after he wrote it. We have ten copies.

When it comes to the New Testament, written between 50-100 A.D, there are more than 5,000 copies. All are within 50-225 years of their original writing. Further, when it came to Scripture, scribes (monks) were meticulous in their copying of original manuscripts. They checked and rechecked their work, to make sure it perfectly matched. What the New Testament writers originally wrote is preserved better than any other ancient manuscript. We can be more certain of what we read about Jesus' life and words, than we are certain of the writings of Caesar, Plato, Aristotle and Homer.

Christians believe that by inspiring the writers' own writing styles and personalities, God shows us who he is and what it's like to know him. There is one central message consistently carried by all 40 writers of the Bible: God, who created us all, desires a relationship with us. He calls us to know him and trust him.[33]

CHAPTER 12.

Is Jesus an actual historical figure, who lived in the Middle East 2000 years ago?

Dabjebrosato, you like many young people may have never heard of Jesus, and wondered who he is and did he really live, or was he only a myth, like the ancient Greek gods

The first evidence that I would present, that Jesus actually lived, is part two of the Bible, known as the New Testament.

Jesus is the central person in the New Testament. In every book, in every chapter, of all that is written, it is about Jesus. It covers his birth, his life, what he taught, his trial, his death by crucifixion, and if it is believed his resurrection.

His resurrection is understood to be, that he came to life again, following being lashed, beaten, hung on a wooden cross for hours, where he finally died. The story goes on to say after He died, His body was taken down, wrapped in burial cloths and laid in tomb, hewn out of a cliff face. A huge round boulder was pushed against the entrance and sealed by Roman soldiers. Then on a morning three days later, some women who were in his group went to the tomb, expecting to embalm his body, but the boulder was found rolled away and the tomb was empty. It was then reported that he first had appeared to one of the women, outside the empty tomb, then to his disciples. Over the next forty days he was seen by, spoke and ate with over five hundred people.

As we have read in chapter 9, of this book, the detail, accuracy and documentation of what is written in the Bible is unsurpassed. I believe that we can be sure that a man named Jesus walked the Earth, in the Country then known as Israel, approximately 2,020 years ago.

Independent evidence.

As well there is I believe one huge often overlooked point that Jesus is a historical person.. This is that the time we live in is divided by his life. By this I mean that for all history, the years before Jesus Christ (his full title) lived, are all dated as B.C. (Or Before Christ). All the years after he lived are dated A.D. (which indicates the Latin words *Anno Domini* which translates in English 'in the year of the Lord). As this dating is on every coin in the world, I would believe it would be impossible for this dating to have occurred if Jesus had not lived.

Which brings me to question 3. Was Jesus actually killed by being crucified (nailed) to a cross?

CHAPTER 13.

Was Jesus actually killed by being crucified (nailed) to a cross?

The first evidence I put forward is from the New Testament, (which as we have seen is the most authenticated book of all time). All four books of the New Testament, known as 'the Gospels' record Jesus' crucifixion in great detail, ending with his death and being placed in a tomb.

Secondly, while there have been several *'theories'* put forward that Jesus did not die on the cross. One being that Jesus only fainted on the cross and was taken down by his disciples and taken away. Another that Jesus' place was taken by another man who died in his place, while Jesus secretly went to another Country.

Whatever else they may claim, at least these *theories* do affirm that Jesus was nailed to a cross.

However looking closely at the *'theories'* I believe that the *'fainting theory'* can be dismissed by the accounts given in the Gospels, that a Roman soldier thrust a spear into Jesus body and a mixture of blood and water came out. The second *'theory'* that someone else died in Jesus place, while he went to another Country, can also be dismissed, by the fact that his disciples told everyone that it was true that Jesus died on the cross and later that he was *resurrected* (returned to life). They did this for the rest of their lives and did not deny it, even when for eleven of the twelve, their death came by violence.

This leads to the next questions. Did Jesus come alive again, after his body was placed in a burial chamber and was he seen (physically) by five hundred people afterwards?

CHAPTER 14.

Did Jesus come alive again, after his body was placed in a burial chamber and was he seen (physically) by five hundred people afterwards?

I believe that the most important part of Jesus' story, is that he was resurrected on the third day after his body was put into a tomb. It is so important because as Christians we believe that the God we worship as well as being our initial Creator, has the power to continue our lives after our earthly bodies die. The importance of this belief was summed up by the New Testament writer Paul, when he wrote in 1 Corinthians Chapter 15 verses 12-19 (precis) If Jesus has not risen from the dead, then we are very mistaken and most to be pitied, as our belief is hopeless..[34]

What is the evidence that Jesus rose from the dead?

From the Bible reading in Matthew Chapter 27. We see after Jesus is pronounced dead, he was taken down from the cross, his body was wrapped in burial clothes and placed in a tomb, which had been hewn out of a rock face. Then a large round shaped rock was rolled across the entrance to block it. As the Sun set, the Jewish Shabbat (Sabbath) day began. On the Shabbat day the

Jewish leaders being worried that Jesus' followers would steal Jesus body away, went to the Roman Governor Pilate and told him of their concerns. So Pilate ordered that a seal be put across the stone blocking the tomb's entrance and a twenty-four-hour guard was put in place, with the orders that they would be put to death, if they allowed anyone to interfere with the stone.

The Bible story continues in Matthew Chapter 28, when at sunrise following Shabbat several of Jesus' women followers went to the Tomb, with aromatic spices to treat Jesus body as it was not done previously because of the haste to get it in the tomb before sunset, when Shabbat started. However to their surprise when they got to the tomb, the guard had disappeared, the seal was broken, the stone was rolled away from the entrance, and the tomb empty. Except the cloth's that had wrapped his body were in two neat piles. The Gospels then give varying accounts of the risen Jesus' (bodily) first meeting with one or more of the women.

What is most important regarding the evidence of women being the first to meet the risen Jesus, is that *'In the first century AD, the testimony of women was not counted as credible and no man in the first century would give credence to a woman's testimony.*[35] So if the disciples were making up the story, they certainly would not make women as being the first witnesses.

The second evidence that the story is true, is that Jesus body has never been found. To discredit the story the Jewish leaders would try their hardest to produce the body. And as we can see by sealing the tomb and putting a guard on it, they did all they could to keep it in the tomb.

The third evidence is that eleven out of the twelve of Jesus' disciples were prepared to die, rather than say it was a lie. It has been said "that a person is willing to die for a lie, that they believe to be true. But a person will not die, for a lie, that they know to be a lie."

So I believe we can be confident that Jesus did return to life, on that first day.

And today as Christians, we believe that He is still in the world in spirit form, (where He is known as the *Holy Spirit).*

Today Jesus' resurrection is remembered on Easter Sunday.

And being present in spirit form, we now look to what makes Christianity unique from other religions? Particularly looking at what are the characteristics of the Christian God and Jesus?

CHAPTER 15.

What are the characteristics of the Christian God and Jesus?

In all other religions the *'God'* is very distant and it is considered impossible for a follower to have a personal relationship with their God, as a son or daughter would with a human father. Such as being able to speak to them personally, asking questions, requesting help, having the feeling of being loved by their *'God'*.

However in Christianity what stands out, is that Christians see their God as having the characteristics of a human father. The overwhelming characteristic (sometimes missing in human fathers) is that of love, in all situations. They feel they can talk with Him (in prayer) in their own words at any time, during every minute of every day. And they believe He is always listening. Although as a Father, He will only grant those requests He knows will be to their good. Sometimes He acts in ways, that they can only see was to their benefit, when they look back over time.

A way that God's love is portrayed is in the story Jesus told of the Prodigal Son in the book of Luke Chapter 15 verses 11-15[36] There Jesus portrays a loving father, as God showing His love for one of the people He has created. *Precis: 'A son had demanded his inheritance from his father. When he got it he went to another*

Country and there wasted the money on loose living. A drought came over the Country and he was reduced to starvation and feeding pigs was all the work he could find. The son realised that he had 'made a big mistake', and he was very sorry, so he made the decision to go home, hoping his father would at least let him be one of his servants. Instead he found his father waiting at the gate, looking for 'his lost son'. His father ran to greet him and forgave him. Put a ring on his finger, along with a robe, and sandals. Then he held a big party, saying "my son who was lost is now found".

When looking at God, it is important to understand that He loves every one of His *created human beings* equally, God has no favourites. A part of that love is wanting each one of the people He created to have the best that life could be and He set out in the Old Testament section of the Bible rules that if followed would keep the people out of trouble.

The most important set of rules is known as the Ten Commandments which can be read in the book of Exodus Chapter 20[37]

As you probably have never heard of them or read them yourself. I will precis them in modern language.

The first one is, I am the Lord your God who brought you out of bondage in Egypt, you shall have NO other gods before me! That includes, being so obsessed, that you worship money, material possessions, being famous, protecting animals or even the environment. You are not to become so busy, thinking and worrying about these things, you forget about me!

Number two. You shall not make any carved or moulded image of a likeness of anything in heaven or on earth and bow down and worship it. For I have feelings just like you, I love you and get very jealous, when you love something else more than you love me!

Three. You shall not use my name as a swear word. *(That now includes Jesus' name).*

Four. Remember the Sabbath (Saturday Jewish, Sunday Christian) day and keep it holy. In six days I created the world and on the seventh day I rested. I want you to do the same, work six days and rest on the seventh. Particularly on the seventh day I want you to think of me and spend time with me. You, your family, foreigners living in your land, everybody, rest and remember me.

Number five. Honour your father and mother, they raised and cared for you when you were growing up, if you don't respect them, who will you respect? Respect is the basis of a good society.

Six. Very straight forward, you shall not deliberately kill or murder someone.

Seven. You shall not have sex with someone who you are not married too. *(With your first sex act you were created to bond, hormonally and psychologically to your husband or wife for life. Sleeping around and viewing pornography breaks this bond and can end in deep regret, broken relationships and broken families).*

Eight. You shall not steal.

Nine. You shall not tell lies about anyone or anything!

Number ten. You shall not want to have your neighbour's house, or wife, or car, or anything else. Be satisfied with what you have.

Well, Dabjebrosato, if everyone tried to follow those rules, the World would be a better place, don't you agree?

And Christians find with experience, that living a life, of trying to follow God's guidance as set out in the Bible, caring for their families and those around them both friends and strangers. Will bring a sense of peace and fulfilment, that does not come from achieving a career and wealth that only benefits yourself and no-one else.

But now we come to the big problem. When God created us, (the human beings in the World) He particularly wanted to have a personal relationship with us, and He couldn't have that relationship if we were zombies with no mind of our own.

So the only way He could have a meaningful relationship with us, was to give us the choice to either be part of that relationship or not. That is called freewill. It is like the difference in the relationship, a person can have with a doll, or a baby. A doll cannot respond in any way while a baby responds with smiles and chuckles and as it gets older he or she can have a relationship of equals with the person.

But with freewill comes the complication, in that as *created human beings* it also allows us to *do our own thing* or *rebel*. And to think we know better than God. That leads to trouble as we will now see.

CHAPTER 16.

Why did Jesus crucifixion become the greatest gift ever?

Now we get to a part that at times will be difficult to understand. As part of that we need to realise that we are trying to understand the actions of an omnipotent God, that we with our small minds, cannot expect to fully understand or comprehend. Particularly as His plans and actions, affect the whole World and an immense period of time.

For Christians it is a time where what is known as *faith* comes in to play. Faith to a Christian *is to believe in something, although you cannot see it and it may not happen until sometime in the future.*

What I would like us, to focus on now, is that God just didn't create us so He could have a relationship here on Earth with us, (important as that can be). But He gave us an immortal soul that Christians believe, lives on in a spiritual form after they die.

That is where Heaven comes into the picture. Christians believe that when they die, they go, in spirit form to live with God in Heaven.

Heaven is where God has told Christians He lives, a place where spirits are restored to full health, it is always peaceful, there are no tears or pain. In Heaven there is a room for each us, but with the requirement that we have to reserve the room and know the password to be able to enter Heaven.

This is where Christians differ greatly, from most other religions and particularly from Atheists.

This because Atheists believe there is no God, only a Universe and World that evolved from nothing, without a purpose, ending in nothing. This also means that for them Heaven cannot exist either.

Although I think for a lot of people, because while they claim not to believe in God, when you look at so many death notices, you get the impression that they still like to hope that when they die they will end up in Heaven anyway. However it will be a shock for them, when they find out there is a Heaven, but they are not able to enter it, because they have not made a booking, or know the password.

But even for those who call themselves Christian entering Heaven may not be a certainty.

This is because of a problem called Sin. Sin is a contaminant, that is created, when we as *'created human beings, with our own freewill'* want to do *our own thing*, and rebel against God and break the rules He has set for us. The result is that we become contaminated by Sin, which the Bible depicts as an *unseen, immoveable stain,* that we can't see, but God can. It is something that happens to everyone, as soon as they are old enough to say "NO, I won't do that!" Or tell lies, or steal something (no matter how small) that doesn't belong to them.

Being contaminated with Sin then creates a problem when it comes to entering Heaven.

Let me explain, because the Christian God has the power that He was able to create the Universe, the World and all that is in it, by just saying a word. He radiates a lot of energy (like a magnetic field force), all around Him and any person who is contaminated with Sin, cannot exist in His presence.

Can I give you an example that you may be able to understand? If you have a burning flame that is very hot like the gas jet on a stove and you bring a clear, pure piece of glass near it, the radiated heat (energy) from the flame will pass straight through it and the glass will be unaffected. On the other hand, if you bring an impure piece of glass, such as one that has been contaminated with a marking pen or black paint, near a burning flame, the heat won't pass through the glass but will quickly heat it and the glass will melt or shatter.

It is the same situation with God and Sin. Sin contaminated people are barred from coming into His presence because they would vaporise in His presence.

It is a real dilemma that began at the beginning of time. The incident is told in the book of Genesis Chapters 2 and 3, in the Old Testament part of the Bible.[38] God had just finished creating the World and He had made a garden which was known as the Garden of Eden.

In the garden He created the first two *human beings* Adam and Eve. They were innocent, in their minds, so they only thought *good thoughts.* God had put an abundance of fruit trees and other food in the garden for Adam and Eve to eat. But He had put one tree in the centre of the garden, which was the *tree of knowledge of good and evil.*

God told Adam and Eve they must not eat the fruit of the *tree of knowledge of good and evil,* because if they did they would die. He said this because He knew if they ate the fruit, they would lose their innocence, become selfish, greedy and rebellious. This would see them becoming contaminated with Sin and as a result they would lose the close relationship they had with Him, they would not be able to remain in God's presence and in the Garden of Eden. Eventually they would die.

The story goes on to say that despite God's warning, Eve and then Adam were tricked by God's arch enemy Satan into eating the fruit (portrayed as an apple) from the *tree of knowledge of good and evil.* Satan, told Eve, God was lying and they could eat the fruit and they would not die!

But the liar was Satan. Adam and Eve's minds were changed, they lost their innocence, they started arguing and blaming each other, for eating the fruit, they realised they were naked and were ashamed, so they made clothes for themselves. The contamination with Sin had begun.

Because of their contamination with Sin, God had to put them out of the Garden of Eden and put a fiery barrier to stop them returning. The Sin contamination became an impossible problem to anyone entering Heaven. This problem of Sin contamination, has been passed down from generation to generation until now

Adam and Eve were in a terrible situation, there was (and still is) no way they could get rid of the Sin. They couldn't see it, in order to deal with it. They couldn't bribe God with money, He has no use for it. Trying to be a good person was no good, because now their minds were contaminated and they continued to make mistakes and think bad thoughts every day. They just could never be good enough.

God still loved Adam and Eve, but how could He clean them of the Sin?

The first solution that God tried to overcome the gap that had developed between him and Man *(His created beings)*, was by focusing on one particular group of people, who became known as the Jews. He made a Covenant, or Agreement, with them that if they would keep his laws, he would be their God and they would be his people.[39]

For the times when they broke his laws, he gave them a series of religious rituals they were to perform, which involved the killing of certain animals (the animals were to be a substitute for the people to sacrifice). The killing of the animals' and the spilling of their blood was required by God as payment (known as atonement) for the people's mistakes and to cleanse them from the stain of their sins. But, this didn't work very well and had to be done over and over again.

Could I say at this stage, the killing of the animals might be shocking to you, but God knows they don't have brains and souls like we do. It is only in recent years that we have put animals on the same level as humans and in some cases, even consider them more important than people. And we need to realise, that, if a death has to occur, God thought it was better it was an animal and not a person.

Which brings us to the extremely important reason of why Jesus allowed himself to be crucified, ending in His death.

About 2,020 years ago, God decided that the sacrifice of animals to cleanse people of their sins, was only a temporary solution and as it had to be done, over and over again. Then the time had come for something permanent to be done and something that would also be available to everyone in the World.

So he decided to make a New Covenant, firstly with his people the Jews and then with the people of all the Nations of the world. As part of this New Covenant, God would provide the ultimate substitute sacrifice, which was to be Jesus. (Yes that is correct, His only son, in human form, who was to be known as *the lamb of God*) and whose death was to replace that of the animals and to offer his *sacrificial death* as gift to every person in the world and for each generation to come.

There was only one requirement: each person had to choose either to believe in and accept Jesus' sacrificial death and accept it as a free gift. (A gift which consists of Jesus symbolically dying in what was to be their place, thereby allowing their souls to continue to live after they died and enter Heaven to be with God). Or to reject it! And personally then remaining contaminated with Sin, facing the consequences of their decision when they died and living in eternity with the result of their choice.

Note that taking one's own life does not avoid, whatever is to happen. It only robs the person of the opportunity, at a later stage of changing their minds and accepting God's gift.

To fulfil his part of the Covenant, God met with Jesus in Heaven and said to him.

"I want you, my Son, to humble yourself, to give up your heavenly rights and to enter the world in the form of a human baby. I want you to grow up and live as a man to show people what

I'm like. Then, at the time I have appointed, I want you to allow yourself to be taken prisoner, be beaten, humiliated, convicted of false charges, and as a consequence be crucified by men and die.

Your death and the spilling of your blood will be the once and for all substitute sacrifice for the sins of every person in the whole world and for all the generations to come.

The only condition is that each person will have to choose, whether to accept your sacrifice, as taking their place or say "No! I will take the consequences myself".

By allowing yourself to be persecuted and suffer, you will show all people that you understand their suffering, as only one who has suffered can.

You will remain dead until the third day and then I will show my supreme power over death by bringing you back to life on earth. After that you will return to me in Heaven".[40]

Jesus was obedient to his Father. His Father's Spirit 'overshadowed' a virgin called Mary, who became pregnant and had a baby, she and her husband Joseph called him Jesus. He grew into a man. He began his ministry by gathering together a group of twelve disciples, then he began teaching all over the country, miraculously healing the blind. disabled, and raising people from the dead. At God's appointed time, he was betrayed, unlawfully convicted and crucified. He died and was placed in a tomb. But on the third day he rose to life again. Then he ascended into Heaven.[41]

God's part of the New Covenant was in place. Jesus' sacrificial death, had replaced the sacrifice of animals, for all the time to come. Jesus offer to die in the place of anyone, in order to pay the price to clean them of their Sin, is the greatest gift that has ever been made.

And as stated previously, there is only one condition to receive it. The person has to acknowledge that they are a Sinner, to be sorry and turn away (try to stop sinning) from their sins, to accept Jesus 'free gift' and then to follow Him as their leader.[42]

Now when a person comes to understand the Christian story, that they are *'created human beings'* or in fact *'God's children but stained with Sin and all the consequences that involves'*. They can understand the wonderful thing that Jesus did. That by dying on the Cross, He made it possible for them to be cleansed of their Sin and the ability to have a close relationship with God and upon their death enter into His presence in Heaven.

And because while He did it for everyone in the World, (who would accept it as a gift), They realise that He did it for them and you and me, even if we were the only person in the World.

Those events as I said happened between 2,020 and 1,990 years ago. We know they happened, because today we remember and celebrate Jesus' birth on Christ-mas Day. And at Easter, we remember Jesus' crucifixion and resurrection.

The question now is, what does all this mean for us today?

CHAPTER 17.

What does all this, mean for us today?

Well to summarise what I believe we need to consider it is this:

Firstly, does the evidence we have looked at show us that we are *created human beings,* and that we have been created for a purpose, in a World created with purpose?

Secondly, does the evidence support the proposition that the *Christian God* is the Omnipotent Creator, who created it all and particularly each of us?

Thirdly from what we have learnt from the Bible, does it show, like it or not, that because we have been born with a freewill and rebellious nature, we have all been contaminated with Sin. When I say all, that includes Presidents, Prime Ministers, Corporate leaders, film stars, musicians, teachers, journalists, sports stars, atheists, everyday men and women, homeless people. Everyone!

Right across the spectrum, from the wealthiest to the poorest, everybody across the World, when they die will have to face the problem of being shut out of Heaven because of being contaminated by Sin. And for a huge number of those people, they do not know, because they have never been told.

Is that fair? I cannot give an answer, I am not God, only He knows.

But for each of us, we have now heard and know we are *created human beings*, with all the implications that means. What will our response be?

We could be like atheists, saying "I do not believe there is a God" and go on living like we are now.

But that is just like a little boy who tells his father that he has an 'invisible cloak'. The little boy puts on the 'invisible cloak' and says to his father, "you can't see me". His father smiles as he sees the boy plainly standing there.

God, can also clearly see the Atheist making this claim, but rather than smiling, He is so sad. Sad because He has offered so much to the Atheist, but the Atheist has rejected Him. This is because the Atheist has failed to see God's hand in His creation, which is all around him and instead believed all the lies he has been told.

Or we could say "this is worth, very serious further investigation, I must look closely at all I have read because if it is true, then I must do all I can to obtain it.

Just the same as the men in Jesus' parable about finding *the kingdom of Heaven*. Both the man who found *the buried treasure*, and the man who found *the pearl of great price*, did everything they could to obtain it. We must do the same.

If you want to investigate further, there is no better place to start than by reading the Bible. In providing the Bible, *God* has given us an instruction manual, which broadly explains how He made it all, and how it should function to get the best results. Particularly in the New Testament, the story of Jesus' life and what He did shows, a way we can live to have the most fulfilling lives possible?

One important thing to know, is that you are in the same position of that of an adopted child. Who when they find out they are adopted, usually have an aching need that is not fulfilled until they have met with their biological father or mother 'face to face.

The exciting similarity in your case, is that you have found out that you have a spiritual Father. And just as the adopted child can seek and find their biological father or mother. You now have the opportunity to seek and find your spiritual Father, to discover His great love for you and to find the answers, that will enable you to cleanse yourself of your Sin, know your self-worth, find purpose and peace in life.

When I talk about peace, I mean freedom from worry, anxiety and depression. God knows this is one of the most important things that affects us, and so He has put the phrase *'fear not' (or a derivation of it)* in the Bible 365 times, or once for everyday of the year.[43]

For Christians it does not mean that all worry, anxiety or depression will be taken away. But it does mean, that the fear of death is diminished as there is the hope of going to Heaven when they die. And all the 'fear nots' are followed by reassurance such as Luke 12:22-26 [22] Jesus said to his disciples, *"Therefore I tell you, don't be anxious for your life, what you will eat, nor yet for your body, what you will wear.* [23] *Life is more than food, and the body is more than clothing.* [24] *Consider the ravens: they don't sow, they don't reap, they have no warehouse or barn, and God feeds them. How much more valuable are you than birds!* [25] *Which of you by being anxious can add a cubit to his height?* ²⁶ *If then you aren't able to do even the least things, why are you anxious about the rest?).* These verses can bring great comfort when fears threaten to become too great.

The decision is now with you. Jesus gave some very good advice in the book of Matthew chapter 7 and verse 7. He said ⁷ *"Ask, and it will be given you. Seek, and you will find. Knock, and it will be opened for you".*⁴⁴

So can I encourage you, to ask in prayer, (talk to Jesus as if you believe He exists today). Seek, (read a chapter of the New Testament each day). Knock, (ask God to reveal himself to you). Try it for a month, I know you will be amazed at what happens, as millions of people have already experienced. I pray that you will find the peace and joy that comes from being able, to pray and know that God is listening, to being able to recognise the answers to those prayers. And having a manual with promises and guidance as how to find the purpose for your life, how to work towards getting the best results possible and that you are not alone in the trials and tribulations that everyone of us will experience.

And I pray that my dream, that I mentioned at the start of this book will have changed. As Nan and I wait inside the Gates of Heaven for you to come, you will arrive. And having accepted Jesus as your Leader and Saviour your booking for a place in Heaven will have been made. You will know the passwords, as you will say "I am with Jesus". The hugs and kisses we will share, at that time, will be worth all the effort it has taken you to get there.

ENDNOTES

1. (Truth definition: A statement proven to be or accepted as true): https://www.thefreedictionary.com/truth

2. The observation, identification, description, experimental investigation, and theoretical explanation of phenomena https://www.thefreedictionary.com/science

3. A philosophical movement of the 1700s that emphasized the use of reason to scrutinize previously accepted doctrines and traditions and that brought about many humanitarian reforms. http://www.thefreedictionary.com/enlightenment

4. A set of statements or principles devised to explain a group of facts or phenomena, especially one that has been repeatedly tested or is widely accepted and can be used to make predictions about natural phenomena. https://www.thefreedictionary.com/theory

5. Dependent upon or characterized by mere chance https://www.thefreedictionary.com/haphazard

6. The text is from https://en.wikipedia.org/wiki/Lottery_mathematics

7. Source Moscow Observer 11/20/2010 https://www.ghanaweb.com/GhanaHomePage/NewsArchive/Evolution-is-mathematically-impossible-197854

8. The text is from Michael J. Behe http://www.arn.org/docs/behe/mb_mm92496.htm

9. The text is By Ker Than Live Science https://www.livescience.com/474-controversy-evolution-works.html

10. All About Science.org http://www.big-bang-theory.com/#sthash.pzuC4kAN.dpuf

11. Science - By Sid Perkins Jul. 17, 2011 http://www.sciencemag.org/news/2011/07/earth-still-retains-much-its-original-heat

12. The text is from https://en.wikipedia.org/wiki/Earth%27s_orbit

13. Earth/Sky Posted by Deanna Conners in Earth | September 22, 2020

    https://earthsky.org/earth/can-you-explain-why-earth-has-four-seasons

14. Text is from Science: By Sid Perkins Jul. 17, 2011 http://www.sciencemag.org/news/2011/07/earth-still-retains-much-its-original-heat

15. The text is from https://en.wikipedia.org/wiki/Primordial_soup

16. Text is from Thoughtco.com By Andrew Zimmerman Jones Updated July 03, 2019

https://www.thoughtco.com/major-laws-of-physics-2699071

17. To cause to exist; bring into being: https://www.thefreedictionary.com/create

18. The text is from https://en.wikipedia.org/wiki/Origin_of_the_Moon

19. The text is from Windows to the Universe http://www.windows2universe.org/the_universe/uts/moon1.html

20. The text is from Thieme https://www.thieme.com/resources/66-resources/resources-for-students/1014-what-can-a-person-survive-the-borders-of-the-human-body

21. The text is From Genetics Home Reference Med Line Plus https://ghr.nlm.nih.gov/primer/basics/dna

22. The text is from Sciencing May 14, 2019 By Dr. Mary Dowd

    https://sciencing.com/would-happen-cell-dna-2424.html

23. The text is from https://en.wikipedia.org/wiki/Microsoft

24. Goodreads quote Bill Gates 'The Road Ahead https://www.goodreads.com/quotes/336336-dna-is-like-a-computer-program-but-far-far-more

25. The text is from https://www.biblegateway.com/passage/?search=Matthew+13&version=WEB

26. 11 Amazing facts about veins by Jordan Rosenfeld February 6, 2018 Mental Floss

    https://www.mentalfloss.com/article/524180/11-amazing-facts-about-veins

27. Exchanging oxygen and carbon dioxide by Rebecca Dezube, MD, MHS, Johns Hopkins University Last full review/revision Jun 2019.https://www.merckmanuals.com/home/lung-and-airway-disorders/biology-of-the-lungs-and-airways/exchanging-oxygen-and-carbon-dioxide

28. https://www.thefreedictionary.com/omnipotent

29. https://www.thefreedictionary.com/faith

30. The text is from Greek Gods and Goddesses. https://greekgodsandgoddesses.net/goddesses/gaea/

31. The text is by Professor Pauline Ladiges AO FAA Professor Kingsley Dixon Dr Anna Koltunow FAA https://www.science.org.au/curious/earth-environment/plant-germination

32. https://www.thefreedictionary.com/prophet

33. The text is from https://www.everystudent.com/features/bible.html

34. https://www.biblegateway.com/passage/?search=1+Corinthians+15%3A12-19&version=WEB

35. The text is by Margaret Manning Shull https://www.rzim.org/read/a-slice-of-infinity/credible-witnesses

36. https://www.biblegateway.com/passage/?search=Luke+15%3A11-32&version=WEB

37. https://www.biblegateway.com/passage/?search=Exodus+20&version=WEB

38. https://www.biblegateway.com/passage/?search=Genesis+2%3A4-3%3A24&version=WEB

39. https://www.biblegateway.com/passage/?search=Exodus+19%3A5%E2%80%936+&version=WEB

40. https://www.biblegateway.com/passage/?search=John+1-21&version=WEB

41. https://www.biblegateway.com/passage/?search=luke+1-24&version=WEB

42. https://www.biblegateway.com/passage/?search=Romans+10%3A9-10&version=WEB

43. The text is by Bill Gaultiere https://www.soulshepherding.org/fear-not-365-days-a-year/

44. https://www.biblegateway.com/passage/?search=matthew+7%3A7&version=WEB

www.ingramcontent.com/pod-product-compliance
Lightning Source LLC
LaVergne TN
LVHW042000060526
838200LV00041B/1795